# 10th International Conference on Biomolecular Engineering (ICBE Asia 2020)



Singapore
7-9 January 2020

ISBN: 978-1-7138-0734-6

**Printed from e-media with permission by:**

Curran Associates, Inc.
57 Morehouse Lane
Red Hook, NY 12571

**Some format issues inherent in the e-media version may also appear in this print version.**



AIChE
120 Wall Street, FL 23
New York, NY 10005-4020

Phone: (800) 242-4363
Fax: (203) 775-5177

www.aiche.org

**Additional copies of this publication are available from:**

Curran Associates, Inc.
57 Morehouse Lane
Red Hook, NY 12571 USA
Phone: 845-758-0400
Fax: 845-758-2633
Email: curran@proceedings.com
Web: www.proceedings.com

# TABLE OF CONTENTS


## PLENARY LECTURE I: PEPTIDE/PROTEIN ENGINEERING

**PLENARY TALK: SUPERGLUE FROM NATURE** ........ 1
*James P. Tam*

## TECHNICAL SESSION 1: ENZYME ENGINEERING

**ENGINEERING THE PLAGUE CAPSULAR ANTIGEN (CAF1) PROTEIN: A THERMALLY REFORMABLE PROTEIN POLYMER** ........ 2
*Daniel T. Peters, Jeremy H. Lakey, Helen Waller, Adrian Yemm, Gema Dura, Neil D. Perkins, David A. Fulton, Mark A. Birch*

**EVALUATING CHEMO-SENSITIZING POTENTIAL OF OXYGEN DELIVERY FACILITATED BY TRANSFUSED POLYMERIZED HEMOGLOBINS ON VASCULARIZED SOLID TUMORS** ........ 3
*Donald Belcher*

**IMPROVING THE PRODUCTIVITY OF 5-HYDROXY-L-TRYPTOPHAN IN ESCHERICHIA COLI BY COMBINATIONAL EVOLUTION OF SEVERAL KEY ENZYMES AND CO-ENZYMES** ........ 4
*Mengjun Fang, Haijiao Wang, Zhinan Xu, Lei Huang*

**INVITED TALK: SPY AND SNOOP PEPTIDE SUPERGLUES TO EMPOWER VACCINES, BIOMATERIALS AND ANTIBODIES** ........ 5
*Mark Howarth*

## TECHNICAL SESSION 2: PROTEIN ENGINEERING AND FORMULATION

**A CLOSER LOOK INTO FREEZE-THAW PROCESSES FOR THERAPEUTIC PROTEIN FORMULATIONS** ........ 6
*Jakob Buecheler*

**AN ENGINEERED SURVIVAL-SELECTION STRATEGY FOR SYNTHETIC BINDING SCAFFOLDS SPECIFICALLY TARGETING POST-TRANSLATIONALLY PHOSPHORYLATED PROTEINS** ........ 7
*Bunyarit Meksiriporn, Matthew P. Delisa, Dujduan Waraho, Allen Jiang, Morgan R. Baltz, Erin Stephens, Hyeon Cheol Lee*

**DIGITAL TWINS OF CONTINUOUS INTEGRATED BIOMANUFACTURING PROCESSES** ........ 9
*Alois Jungbauer*

**SITE-SPECIFIC MODIFICATION OF IGG VIA A SIMPLE PHOTOCROSSLINKING REACTION** ........ 10
*Jisoo Park, Tae Hyeon Yoo, Jeung Yeop Shim*

## PLENARY LECTURE II: BIOMOLECULAR ENGINEERING & SELF-ASSEMBLY

**PLENARY TALK: "BIOMIMETIC STRUCTURES FOR REGENERATIVE MEDICINE"** ........ 11
*Samuel Stupp*

## RAPID FIRE POSTER PRESENTATIONS

**DISPLAY OF FUNCTIONAL NUCLEIC ACID POLYMERASE ON ESCHERICHIA COLI SURFACE AND ITS APPLICATION IN DIRECTED POLYMERASE EVOLUTION** ........ 12
*Muen Chung*

EFFECT OF NUTRIENT SUPPLEMENT ON SYNGAS FERMENTATION WITH CLOSTRIDIUM AUTOETHANOGENUM ..... 13
*Young-Kee Kim, Hien Nguyen Thi*

## TECHNICAL SESSION 3: CELLULAR ENGINEERING

INTEGRATING ENZYME AND METABOLIC ENGINEERING FOR APOCAROTENOIDS PRODUCTION ..... 14
*Xixian Chen, Sudha Shukal, Congqiang Zhang*
INVITED TALK: IN VITRO SYNTHETIC BIOLOGY USING CELL FREE SYSTEMS TO PROTOTYPE PARTS AND PATHWAYS TO ENZYMATIC CONVERSION IN A TEST-TUBE ..... 15
*Paul S. Freemont*
SYNTHETIC GENETIC DEVICES FOR DECODING AND ENCODING CHEMICAL SIGNALS IN LIVING BACTERIAL CELLS ..... 16
*Kathakali Sarkar, Sangram Bagh*
SYNTHETIC GENETIC DEVICES FOR HIGHER-ORDER ARTIFICIAL SIGNAL PROCESSING AND COMPUTATION IN LIVING CELLS ..... 17
*Deepro Bonnerjee, Sangram Bagh*

## TECHNICAL SESSION 4: SELF-ASSEMBLY

BIOENGINEERING PROKARYOTIC NANOCOMPARTMENTS INTO PHOTOSENSITIZING NANOPARTICLES ..... 18
*Andrew Care, Dennis Diaz, Anwar Sunna, Xavier Vidal Asensio*
LIPID-PEPTIDE CO-ASSEMBLIES FOR MULTIFUNCTIONAL BIOINTERFACES ..... 19
*Stefan Salentinig, Mahsa Zabara, Mark Gontsarik*

## PLENARY LECTURE III: BIOMOLECULAR ANALYTICS AND CHARACTERIZATIONS

PLENARY TALK: CELL SURFACE TOPOGRAPHY: HOW PROTEIN SIZE CAN ALTER ORGANIZATION AND SIGNALING AT CELL-CELL INTERFACES ..... 21
*Daniel A Fletcher*

## TECHNICAL SESSION 5: ANALYTICS

EXPLORE THE ADHESION MECHANISMS OF MUSSEL ADHESIVE PROTEINS ..... 22
*Jing Yu*
INVITED TALK: ZIKA VIRUS AND ITS ANTIBODY COMPLEX STRUCTURES ..... 23
*Shee-Mei Lok*
TIME-RESOLVED SMALL-ANGLE X-RAY SCATTERING FOR MONITORING THE BIOMOLECULAR ASSEMBLY PROCESS ..... 24
*Masamichi Ikeguchi*

## TECHNICAL SESSION 6: APPLICATIONS

CONVERTING FATTY ACIDS TO ALKANES WITH BLUE LIGHT ..... 25
*Julie Lake, Samuel Curran*
IP IS DECREASED MELANOGENESIS USING INHIBITION OF PKA ..... 26
*Hwa Jun Cha*

## POSTER SESSION

"PARALLEL METABOLIC PATHWAY ENGINEERING" FOR THE BIOPRODUCTION OF SHIKIMATE PATHWAY DERIVATIVES ..... 27
*Ryosuke Fujiwara, Tsutomu Tanaka, Akihiko Kondo, Shuhei Noda*

**CXCR4 REGULATES BRANCHING MORPHOGENESIS IN DEVELOPING EPITHELIAL ORGAN** ..... 28
*Junchul Kim*
**DEVELOPMENT OF COMPUTER-AIDED RATIONAL PROTEIN ENGINEERING TOOLKIT (CARPET)** ..... 29
*Hyeon Nae Jeon, Hocheol Lim, Sungbo Hwang, Jongwan Kim, Kyoung Tai No*
**DIFFERENCES IN LOOP MOTIONS AND DYNAMICS IN THE CATALYTIC DOMAINS OF GH FAMILY 19 CHITINASES REVEALED BY MD SIMULATION** ..... 30
*Ayokunmi Oyeleye, Adam Thean Chor Leow, Noor Baity Saidi, Normi Mohd Yahaya*
**METABOLIC ENGINEERING TO IMPROVE 1,5-DIAMINOPENTANE PRODUCTION FROM CELLOBIOSE USING BGL-SECRETING CORYNEBACTERIUM GLUTAMICUM** ..... 31
*Akihiko Kondo, Tsutomu Tanaka*
**MICROBIAL PLATFORM TO SYNTHESIZE VALUABLE BULK CHEMICALS IN ESCHERICHIA COLI** ..... 32
*Shuhei Noda*
**POLARISOME SCAFFOLDER SPA2-MEDIATED MACROMOLECULAR CONDENSATION OF AIP5 FOR ACTIN POLYMERIZATION** ..... 33
*Ying Xie, Jialin Sun, Xiao Han, Alma T. Wunder, Yonggui Gao, Yansong Miao*
**SALT-INDUCED ASSEMBLY KINETICS OF ARCHAEOGLOBUS FULGIDUS FERRITIN MONITORED BY TIME-RESOLVED SMALL-ANGLE X-RAY SCATTERING** ..... 34
*Takumi Kuwata, Daisuke Şato, Sierin Lim, Masamichi Ikeguchi, Ambrish Kumar, Boyce Law, Tabitha Tan*
**SELECTION OF PEPTIDES INHIBITING P53 FUNCTION** ..... 35
*Farid Ghadessy, Sharon Chee, Siau Jia Wei, Lim Ting Xiang, Jimmy Yen-Chu Lin*
**SOLAR SPECTRUM CHARACTERIZATION, IMPACT ON HUMAN HEALTH AND MITIGATION USING INNOVATIVE MATERIALS** ..... 36
*Dima Abu Alhawa, Laya Roustazadeh, Konstantinos Kokosimos, Amani Grati, Asma Abousrafa*
**STUDYING PERFORMANCE OF MICROBIAL FUEL CELL INOCULATED WITH SHEWANELLA ONEIDENSIS MR-1 FOR COPPER RECOVERY** ..... 37
*Arjun Bhagwat*
**THE COMPARISON OF MSC SECRETOME EFFECTS ON CORNEAL WOUND HEALING DEPENDING ON THE CELL CULTURE ENVIRONMENTS** ..... 38
*Hyun Jong Lee*
**THEASAPONIN E1 ATTENUATE AMYLOID BETA(A?) PROTEINS AND HYPERPHOSPHORYLATED TAU IN SWEAPP N2A AND SHY-5Y CELLS BY DOWN OR UP-REGULATION OF THE ASSOCIATED SIGNALING MOLECULES AND ENZYMES** ..... 39
*Muhammad Imran Khan, Jong Deog Kim, Min Yong Kim*
**Author Index**

# Plenary Talk: Superglue from Nature

**Conference:** International Conference Biomolecular Engineering ICBE

**Year:** 2020

**Proceeding:** ICBE Asia 2020 - 10th International Conference on Biomolecular Engineering

**Group:** General Submissions

**Session:** Plenary Lecture I: Peptide/Protein Engineering

**Time:** Tuesday, January 7, 2020 - 9:00am-9:25am

## Authors:

Tam, J. P., Nanyang Technological University

Over the past 25 years, major advances are ligation chemistries to form peptide or non-peptide bonds between molecules. These advances are enabled by the development of novel chemical ligation methods and expansion of genetic codons as well as the discovery of inteins and peptide ligases. Of particular interest to our laboratory are the discoveries of plant-derived ligases, which are superglue from nature. These ligases or superglue enable bonding between peptides, peptide-to-protein, and protein-to-protein, without protecting groups, activating agents and under aqueous conditions, and with exquisite selectivity. Here, I will present our work in ligation chemistry to enable site-specific bonding of chemicals, polymers, peptides and proteins to form new compounds under physiological conditions. In particular, I will discuss our work on the Asx-specific peptide ligases such as the butelases, which were discovered in NTU campus and which act as superglue for labeling proteins and live cells as well as precision biomanufacturing industrial enzymes and therapeutics under environmental friendly conditions.

# Engineering the Plague Capsular Antigen (Caf1) Protein: A Thermally Reformable Protein Polymer

**Conference:** International Conference Biomolecular Engineering ICBE

**Year:** 2020

**Proceeding:** ICBE Asia 2020 - 10th International Conference on Biomolecular Engineering

**Group:** General Submissions

**Session:** Technical Session 1: Enzyme Engineering

**Time:** Tuesday, January 7, 2020 - 10:55am-11:20am

## Authors:

Peters, D. T., Newcastle University
Lakey, J. H., Newcastle University
Waller, H., Newcastle University
Yemm, A., Newcastle University
Dura, G., Newcastle University
Perkins, N. D., Newcastle University
Fulton, D. A., Newcastle University
Birch, M. A., Cambridge University

Caf1 is a proteinaceous polymer produced by *Yersinia pestis* bacteria that helps them to evade phagocytosis. Individual 15 kDa Caf1 subunits associate non-covalently to form long (MDa, Âµm), highly thermostable polymers. Caf1 polymers can be combined with cross-linkers to form hydrogels of tuneable porosity and stiffness, are free from animal products, and are simple to produce and purify. Additionally, cells adhere poorly to the material, making it â€œnon-stickâ€. Through mutagenesis, this phenotype can be reversed, allowing different bioactivities to be engineered into the protein. These properties make Caf1 a uniquely interesting biomaterial to develop for biomedical applications.

Here, we engineer Caf1 mutants containing signals involved in osteogenesis, and combine these to form a â€œmosaicâ€ Caf1 polymer that can trigger the early stages of bone formation when used as a scaffold for bone marrow stromal cells. We next demonstrate that denatured Caf1 polymers can self-assemble into oligomers, including mosaic oligomers, allowing us to control the composition of bioactive subunits within the material. Finally, we show that we can use this property to create thermally reformable Caf1 hydrogels, providing a simple and effective route towards cell encapsulation within the 3D material. We anticipate that these developments will increase Caf1â€™s utility as a biomaterial for use in 3D cell culture, as well as other applications.

# Evaluating Chemo-Sensitizing Potential of Oxygen Delivery Facilitated By Transfused Polymerized Hemoglobins on Vascularized Solid Tumors

**Conference:** International Conference Biomolecular Engineering ICBE

**Year:** 2020

**Proceeding:** ICBE Asia 2020 - 10th International Conference on Biomolecular Engineering

**Group:** General Submissions

**Session:** Technical Session 1: Enzyme Engineering

**Time:** Tuesday, January 7, 2020 - 11:20am-11:45am

## Authors:

Belcher, D., The Ohio State University

A major constraint in chemo-and radio-therapeutic cancer treatment is inadequate oxygenation of solid tumors. Consequently, alleviating hypoxia in solid tumors is considered a promising target for improving the efficacy of anti-cancer therapeutics such as chemotherapy. Our group has synthesized variable molecular weight (MW) Polymerized human hemoglobin (PolyhHb) with low (T-state) and high (R-state) oxygen ($O_2$) affinities for use as RBC substitutes. These $O_2$ carriers can be transfused to increase solid tumor oxygenation and improve the efficacy of anti-cancer therapeutics. In this study, we analyzed the biophysical properties of the synthesized PolyhHbs. Three-dimensional computational models of blood flow and $O_2$ transport in the tumor micro-environment were used to examine how dosage and type of PolyhHb impacted tumor $O_2$ delivery. To compliment this study, an animal study was performed by periodically delivering PolyhHb in the T- and R-state. Decreases in the apparent viscosity resulting from PolyhHb exchange transfusion may result in significant changes in flow distributions throughout the tumor micro-circulatory network. The difference in wall shear stress implies that PolyhHb may have a more significant effect on capillary beds. The increased $O_2$ flux and decreased $pO_2$ drop per unit length indicates that both PolyhHbs are suited to deliver $O_2$ under hypoxic conditions. Both T- and R-state PolyhHb exchange transfusion may lead to elevated $O_2$ delivery at low $pO_{2,in}$. In the experimental mouse model, HBOC transfusion led to decreased angiogenesis, tumor growth, and hypoxic gene expression.

# Improving the Productivity of 5-Hydroxy-L-Tryptophan in Escherichia coli By Combinational Evolution of Several Key Enzymes and Co-Enzymes

**Conference:** International Conference Biomolecular Engineering ICBE

**Year:** 2020

**Proceeding:** ICBE Asia 2020 - 10th International Conference on Biomolecular Engineering

**Group:** General Submissions

**Session:** Technical Session 1: Enzyme Engineering

**Time:** Tuesday, January 7, 2020 - 11:45am-12:10pm

## Authors:

Fang, M., ZheJiang University
Wang, H., ZheJiang University
Xu, Z., ZheJiang University
Huang, L., ZheJiang University

In our previous work, one combined biosynthetic pathway was engineered to produce 5-hydroxy-L-tryptophan (5-HTP) with glycerol as carbon source in *E. coli*. In the present work, the low specific bioactivity of tryptophan hydroxylase and low stability of two co-enzymes (pterin-4Î±-carbinolamine dehydratase (PCD) and dihydropteridine reductase (DHPR) ) were identified to be two factors which seriously hinders further enhancement of 5-HTP production. A gene-circuit based high-throughput screening method was developed to evolve the above proteins. One mutant of TPH-8 was screened out to have 1.5-fold bioactivity improvement, and the mutants of PCD 94 or DHPR 109 were also obtained to have obviously improved stabilities. By combining the evolution of all the three proteins, the titer of 5-HTP was improved 3.7 fold compared with the original strain in the shake flask, and no L-trp was left which would make the purification of 5_HTP from the culture medium much easier . The present work would pave one new road for the bioproduction of 5-hydroxy-L-tryptophan in industry.

# Invited Talk: Spy and Snoop Peptide Superglues to Empower Vaccines, Biomaterials and Antibodies

**Conference:** International Conference Biomolecular Engineering ICBE

**Year:** 2020

**Proceeding:** ICBE Asia 2020 - 10th International Conference on Biomolecular Engineering

**Group:** General Submissions

**Session:** Technical Session 1: Enzyme Engineering

**Time:** Tuesday, January 7, 2020 - 10:30am-10:55am

## Authors:

Howarth, M., University of Oxford

A special feature of *Streptococcus pyogenes* enables spontaneous isopeptide bond formation within certain cell-surface proteins. We engineered this system to generate an irreversible peptide-protein interaction (SpyTag/SpyCatcher). This system is rapid, genetically-encodable and specific in diverse biological environments. Latest advances include accelerating to infinite affinity and a toolbox of modules for rapidly controlling protein architectures. Cyclizing enzymes using SpyTag conferred resilience to boiling, for biotransformation and nutrition. SpyTag and the related superglue SnoopTag allow programmable synthesis of multi-functional teams or biomaterials, to modulate precisely cancer cell signaling. Virus-like particles (VLPs) are nano-assemblies with many attractive features for vaccination. However, decorating VLPs with antigens by genetic fusion or chemical modification is often unsuccessful. We demonstrated complete reaction to SpyCatcher-VLPs after mixing with SpyTag linked to a range of malaria antigens and cancer targets. Spy-VLPs efficiently induced antibody responses after only a single immunization and in the absence of adjuvant. Modular assembly using SpyTag should accelerate vaccine development against a range of human and veterinary diseases.

# A Closer Look into Freeze-Thaw Processes for Therapeutic Protein Formulations

**Conference:** International Conference Biomolecular Engineering ICBE

**Year:** 2020

**Proceeding:** ICBE Asia 2020 - 10th International Conference on Biomolecular Engineering

**Group:** General Submissions

**Session:** Technical Session 2: Protein Engineering and Formulation

**Time:** Tuesday, January 7, 2020 - 2:30pm-3:00pm

## Authors:

Buecheler, J., Novartis

Freezing and thawing are essential processes of the life cycle for a therapeutic protein. They enable stability during transport or storage as well as decoupling of DS and DP shelf-life for most Biologics Bulk Drug Substances (BDS).

Reports and publications about the impact of freezing and thawing on the stability of therapeutic proteins are contradictory. Some link a fast freezing process with an increase in protein degradation, others report the reverse.

Protein stress factors during freezing include an increase in interfaces (such as ice to liquid or to air), but also cold denaturation or cryoconcentration could lead to degradation. Furthermore, it has been reported that cryo-protectants, which should stabilize the protein in frozen state, can crystalize during frozen storage and can cause aggregate formation.

The importance of the thawing process is often being underestimated, as a 300% increase in protein concentration is not uncommon due to stratification effects and can cause degradation. Additionally, various studies show that slow and apparent mild thawing conditions at low temperatures can lead to gelation or precipitation due to protein-protein interactions.

# An Engineered Survival-Selection Strategy for Synthetic Binding Scaffolds Specifically Targeting Post-Translationally Phosphorylated Proteins

**Conference:** International Conference Biomolecular Engineering ICBE

**Year:** 2020

**Proceeding:** ICBE Asia 2020 - 10th International Conference on Biomolecular Engineering

**Group:** General Submissions

**Session:** Technical Session 2: Protein Engineering and Formulation

**Time:** Tuesday, January 7, 2020 - 3:00pm-3:20pm

## Authors:

Meksiriporn, B., King Mongkut's Institute of Technology Ladkrabang
DeLisa, M. P., Cornell University
Waraho, D., King Monkut's University of Technology Thonburi
Jiang, A., Cornell University
Baltz, M. R., Cornell University
Stephens, E., Cornell University
Lee, H. C., ForBioKorea co, ltd

Protein phosphorylation plays an important role in the regulation of protein function and many cellular processes. Aberrant phosphorylation has been shown to be a cause of cell death as well as maligination. As such, there is an urgent need for affinity reagents that target phospho-modified sites on individual proteins for either immunodetection or immunotherapy; however, generating such reagents remains a significant challenge. Here, we described a genetic selection strategy termed PhLI-TRAP (*ph*ospho-functional *l*igand-binding *i*dentification by *T*at-based*r*ecognition of *a*ssociating *p*roteins) for routine laboratory isolation of phospho-specific designed ankyrin repeat proteins (DARPins) by linking in vivo affinity capture of a phosphorylated target protein with antibiotic resistance of *Escherichia coli* cells. The assay is validated using an existing panel of DARPins that selectively bind the nonphosphorylated (inactive) form of extracellular signal-regulated kinase 2 (ERK2) or its doubly phosphorylated (active) form (pERK2). PhLI-TRAP was successfully used to improve the affinity of a phospho-specific DARPin for its cognate pERK2 antigen as well as to reprogram the specificity of the same parental DARPin for binding to noncognate ERK2. Importantly, by linking antibiotic resistance with phospho-epitope binding in the cytoplasm of E. coli cells, the PhLI-TRAP method eliminates the need for purification or immobilization of the phosphoprotein target and only requires selective plating of bacteria on solid medium to uncover productive binders. Collectively, these results establish our genetic selection as a useful and potentially generalizable protein engineering tool for studying phospho-specific binding proteins and customizing their affinity and selectivity.

# Digital Twins of Continuous Integrated Biomanufacturing Processes

**Conference:** International Conference Biomolecular Engineering ICBE

**Year:** 2020

**Proceeding:** ICBE Asia 2020 - 10th International Conference on Biomolecular Engineering

**Group:** General Submissions

**Session:** Technical Session 2: Protein Engineering and Formulation

**Time:** Tuesday, January 7, 2020 - 2:00pm-2:30pm

## Authors:

Jungbauer, A., BOKU

A lot of industries already realized Industry 4.0 and Industry 5.0 technologies into their manufacturing. This enables smart automation and model-based control strategies. In biopharmaceutical industry digitalization is still not common. The product is mainly performed in batchwise manner and the focus is on in-process and release analytics, which are executed off-line and time consuming. The imminent potential for process optimization control and predictive maintenance cannot be fully utilized without manufacturing in continuous mode and a digital representations of bioprocesses. Models for bioprocesses can be based on understanding of underlying biophysical mechanism e.g. heat/mass balance, adsorption/desorption kinetics or steric hindrance or similar and combined with computational fluid dynamics. Such mechanistic models are available to describe and predict chromatography, filtration, and centrifugation. Strategies will be shown how to convert a batch into a continuous integrated bioprocess and the benefits and challenges will be discussed such as reduction of floor space, high productivity but also the problem of propagation of process disturbances will be addressed. For a continuous end to end antibody manufacturing process the residence time distribution model will be shown and how process disturbance propagate. Finally, a roadmap will be presented how we could arrive at a digitized or virtual bioprocess.

# Site-Specific Modification of IgG Via a Simple Photocrosslinking Reaction

**Conference:** International Conference Biomolecular Engineering ICBE

**Year:** 2020

**Proceeding:** ICBE Asia 2020 - 10th International Conference on Biomolecular Engineering

**Group:** General Submissions

**Session:** Technical Session 2: Protein Engineering and Formulation

**Time:** Tuesday, January 7, 2020 - 3:20pm-3:40pm

## Authors:

Yoo, T. H., Ajou University
Park, J., Ajou University
Shim, J. Y., Noroo Holdings Co., Ltd.

Conjugation of antibody has expanded its applications in therapeutics and diagnostics, and various methods have been developed based on chemical or enzymatic reactions. However, majority of them have focused on synthetic molecules such as small molecules, nucleic acids, or synthetic materials, but site-specific conjugation of antibody with protein cargos has been rarely demonstrated so far. In this presentation, I describe a PEptide-DIrected Photocrosslinking (PEDIP) reaction for site-specific conjugation of IgG with protein using an Fc-binding peptide and a photoreactive amino acid analogue, and demonstrate this method by developing an immunotoxin composed of a Her2-targeting IgG (trastuzumab) and an engineered Pseudomonas exotoxin A (PE24). The ADP-ribosylation of eukaryotic elongation factor-2 by the bacterial toxin inhibited the ribosomal translation of protein, and the trastuzumab-PE24 conjugate exhibited the cytotoxicity toward Her2-overexpressing cell lines. The PEDIP reaction can also be applied for many other types of cargo with slight modifications of the method.

# Plenary Talk: "Biomimetic Structures for Regenerative Medicine"

**Conference:** International Conference Biomolecular Engineering ICBE

**Year:** 2020

**Proceeding:** ICBE Asia 2020 - 10th International Conference on Biomolecular Engineering

**Group:** General Submissions

**Session:** Plenary Lecture II: Biomolecular Engineering & Self-assembly

**Time:** Wednesday, January 8, 2020 - 1:30pm-1:55pm

## Authors:

Stupp, S., Northwestern University

Due to shifting demographics, one of the grand challenges for science in this century is to create strategies to regenerate parts of the human body in order to achieve longer â€œhealthspansâ€. Ideally, designed bioactive materials could act as extracellular matrices with finite half-life that traffic signals in dynamic fashion. This in fact would mimic the function of natural matrices as tissues develop or repair after injury, and requires molecular design of soft materials to directly activate signaling pathways. An important feature of the ideal biomaterials is the nature of internal dynamics across scales, thus imitating the non-static nature of living matter. This lecture will describe a broad platform of *supramolecular biomaterials* built with a toolbox self-assembling peptide amphiphiles, glycans, and nucleic acids that has been validated by pre-clinical models of neural and musculoskeletal regeneration. These systems can be crafted as highly effective growth factor mimics with longer half lives than proteins, and the tunable dynamics of their non-covalently bonded molecules can have a profound effect on biological efficacy. The lecture will also describe systems in which biological signals can be switched on and off through external cues, and others in which reversible self-assembly of superstructures modulates the phenotype of brain cells.

# Display of Functional Nucleic Acid Polymerase on Escherichia coli Surface and Its Application in Directed Polymerase Evolution

**Conference:** International Conference Biomolecular Engineering ICBE

**Year:** 2020

**Proceeding:** ICBE Asia 2020 - 10th International Conference on Biomolecular Engineering

**Group:** General Submissions

**Session:** Rapid Fire Poster Presentations

**Time:** Wednesday, January 8, 2020 - 11:30am-11:55am

## Authors:

Chung, M., SeSaM-Biotech

DNA polymerase-based techniques such as PCR and DNA sequencing have become indispensable for modern molecular biology. These techniques foster an expanding field of applications, including clinical diagnostics, drug screening, forensic research and archaeology. To tailor the properties of a polymerase for the desired application, various directed evolution strategies have been employed, such as phage display, compartmentalized self replication and primer extension assay. Here we report, to the best of our knowledge, the first high-throughput screening method for DNA polymerase based on cell surface display (CSD). CSD separates the polymerase from cytosolic materials which might interfere with the selection of the polymerase for desired activities. A DNA polymerase is displayed on *Escherichia coli (E.coli)* cell surface using an inactivated esterase autotransporter. The activity of the displayed polymerase is then detected by a dye that intercalates into elongated DNA and emits more fluorescence. Using this method, we evolved Klenow fragment of *E. coli* DNA polymerase I to incorporate NTP and C2'-OMe modified oligonucleotides. The CSD-based screening method provides a novel alternative for directed evolution of DNA polymerase.

# Effect of Nutrient Supplement on Syngas Fermentation With Clostridium Autoethanogenum

**Conference:** International Conference Biomolecular Engineering ICBE

**Year:** 2020

**Proceeding:** ICBE Asia 2020 - 10th International Conference on Biomolecular Engineering

**Group:** General Submissions

**Session:** Rapid Fire Poster Presentations

**Time:** Wednesday, January 8, 2020 - 11:55am-12:20pm

## Authors:

Kim, Y. K., Hankyong National University
Thi, H. N., Hankyong National University

The syngas fermentation process is attracting attention as a technology to produce bioethanol from lignocellulosic biomass based syngas or waste gas, but feasibility improvement is required for industrial utilization. The cost of the culture media required for fermentation is a big part of overall fermentation cost, and the yeast extract, which is widely used as nutrient supplement, is a relatively expensive ingredient used to make basal medium for cultivation of acetogenic bacteria. In this study, we investigated the effect of corn steep liquor, malt extract, and vegetable extract as a low cost nutrient substitute for yeast extract on microbial growth and product formation during syngas fermentation. We verified that corn steep liquor, malt extract, and vegetable extract could replace yeast extract in syngas fermentation. Moreover, the use of malt extract and vegetable extract resulted in a significant improvement in ethanol productivity.

# Integrating Enzyme and Metabolic Engineering for Apocarotenoids Production

**Conference:** International Conference Biomolecular Engineering ICBE

**Year:** 2020

**Proceeding:** ICBE Asia 2020 - 10th International Conference on Biomolecular Engineering

**Group:** General Submissions

**Session:** Technical Session 3: Cellular Engineering

**Time:** Wednesday, January 8, 2020 - 9:55am-10:20am

## Authors:

Chen, X., BioTrans, A*STAR Singapore
Shukal, S., BioTrans, A*STAR Singapore
Zhang, C., BioTrans, A*STAR Singapore

Metabolic engineering aims to balance intracellular pathways and increase the precursor supply. However, some heterologous enzymes are not evolved to support high flux. To remove the limitation, the catalytic properties of rate-limiting enzymes must be enhanced. Previously, we have demonstrated the use of fusion-partners to improve the expression of carotenoid cleavage dioxygenase 1 (CCD1) in *Escherichia coli*. Here, we further engineered CCD1 whose intrinsic promiscuity and low activity limited the production of Î±-ionone in *Escherichia coli*. Site-directed mutagenesis was carried out to mutate three structural elements of CCD1: an active site loop, Î·-helices and Î±-helices. Furthermore, mutated CCD1 was fused with lycopene epsilon-cyclase to facilitate substrate channelling. Collectively, these methods improved Î±-ionone concentration by > 2.5-fold as compared to our previously optimized strain. Lastly, the engineered enzyme was used in conjunction with metabolic engineering strategy to further boost Î±-ionone concentration by another 20%. This work deepens our understanding of CCD1 catalytic properties and proves integrating enzyme and metabolic engineering can be synergistic for higher microbial production yield.

(1) Zhang, C.; Chen, X.; Lindley, N. D.; Too, H.-P. A â€œPlug-n-Playâ€ Modular Metabolic System for the Production of Apocarotenoids.*Biotechnology and Bioengineering* **2018**, *115* (1), 174â€“183.https://doi.org/10.1002/bit.26462.

(2) Chen, X.; Shukal, S.; Zhang, C. Integrating Enzyme and Metabolic Engineering Tools for Enhanced Î±-Ionone Production. *J. Agric. Food Chem.* **2019**. https://doi.org/10.1021/acs.jafc.9b00860.

# Invited Talk: In Vitro Synthetic Biology Using Cell Free Systems to Prototype Parts and Pathways to Enzymatic Conversion in a Test-Tube

**Conference:** International Conference Biomolecular Engineering ICBE

**Year:** 2020

**Proceeding:** ICBE Asia 2020 - 10th International Conference on Biomolecular Engineering

**Group:** General Submissions

**Session:** Technical Session 3: Cellular Engineering

**Time:** Wednesday, January 8, 2020 - 9:30am-9:55am

## Authors:

Freemont, P. S., Imperial College London

Cell-free transcription/translation systems (known as CFPS or TX-TL) have recently been re-evaluated as a promising platform for enabling synthetic biology research and applications. In particular CFPS has been shown to provide a reproducible prototyping platform for regulatory elements where measurements in vitro are in part consistent with similar measurements in vivo. The advantage of being non-GMO allows rapid automated assays for characterizing parts and genetic circuit designs for pathway engineering, natural product discovery and biosensor designs and field implementation. My lab has been interested in exploring cell free extracts from non-model organisms and I will present our most recent work on cell-free extract systems. I will also present our recent studies on combining in vitro geno-chemtic strategies to allow rapid access to xenobiotic compounds which may provide improved therapeutic activity. By focusing on the violacein biosynthesis pathway and using seven different substrate analogues, we have been able to generate new to nature analogues of violacein. Further new derivatives were also generated from brominated analogues via Suzuki-Miyaura cross-coupling reaction directly using the crude extract without prior purification. This approach shows that biosynthesized natural products can be chemically derivatized in cell extracts.

# Synthetic Genetic Devices for Decoding and Encoding Chemical Signals in Living Bacterial Cells

**Conference:** International Conference Biomolecular Engineering ICBE

**Year:** 2020

**Proceeding:** ICBE Asia 2020 - 10th International Conference on Biomolecular Engineering

**Group:** General Submissions

**Session:** Technical Session 3: Cellular Engineering

**Time:** Wednesday, January 8, 2020 - 10:20am-10:45am

## Authors:

Sarkar, K., Homi Bhabha National Institute
Bagh, S., Saha Institute of Nuclear Physics

Forward engineering of synthetic gene circuits has made higher order information processing feasible to be performed in living cells. One of the next challenges in Synthetic biology is to create artificial decision-making signal processing system with more complex human-defined function and real world application. Here, we develop biological genetic circuits which are analogous to electronic 2-to-4 decoder and 4-to-2 priority binary encoder at the population level in living *E. coli* using synthetic logic gates, regulated by synthetic promoters containing binding sites for several transcription factors (TFs). We consider extracellular chemicals as inputs such as Isopropyl Î²-D-1thiogalactopyranoside (IPTG), anhydrotetracycline (aTc), n-acyl homoserine lactone (AHL) & arabinose and various fluorescent proteins as outputs. We have also used one mutated version of Î» repressor CI: frame-shifted CI which we built and characterized in our lab. We have designed the component circuits based on the electronic decoder and encoder circuit conventions, constructed and characterized to get expected logic behaviors, studied their mathematically predictable properties and established agreement between their mathematical models and experimental outputs in terms of fluorescent protein expression.

# Synthetic Genetic Devices for Higher-Order Artificial Signal Processing and Computation in Living Cells

**Conference:** International Conference Biomolecular Engineering ICBE

**Year:** 2020

**Proceeding:** ICBE Asia 2020 - 10th International Conference on Biomolecular Engineering

**Group:** General Submissions

**Session:** Technical Session 3: Cellular Engineering

**Time:** Wednesday, January 8, 2020 - 10:45am-11:10am

## Authors:

Bonnerjee, D., Homi Bhabha National Institute
Bagh, S., Saha Institute of Nuclear Physics

Biology is a complex arena that involves a multitude of inputs and outputs occurring at the right levels, time and combination to achieve the fine tuning called life. These interaction networks implement robust genetic circuitries created over time. Today, Synthetic Biology demands Synthetic Genetic networks that can function involving larger sets of inputs and outputs and perform more complex decision-making regimes. Despite the plethora of synthetic circuits already designed, we are still far behind the genetic complexity of nature and thus far from attaining the full potential of engineering genomes and cellular robotics. Towards this, we have been trying to develop a few complex genetic computational circuits- namely a biological 1-2 multiplexer (input selector) and a 2-1 de-multiplexer (output selector) amongst others- conceived from electronic circuit design principles. Attaining desired regulation through external chemical inducers, transcription factors and translational control, these circuits have required designing and characterizing libraries of synthetic Promoters and RBS's. Fine tuning of the transcriptional- translational regulatory machinery has culminated in satisfactory Boolean response. Furthermore, we have also explored how minimalistically can one increase the input-output complexities of a synthetic gene circuit through developing a 3 input-3 output combinatorial genetic system. It orthogonally senses three small molecule signals as inputs and responds through three fluorescent proteins in accordance with input logical states. Consensus of the Mathematical models and experimental validations of the systems show good predictability. These circuits may be integrated in future into larger circuits and are stepping stones to more complex application-oriented genetic networks.

# Bioengineering Prokaryotic Nanocompartments into Photosensitizing Nanoparticles

**Conference:** International Conference Biomolecular Engineering ICBE

**Year:** 2020

**Proceeding:** ICBE Asia 2020 - 10th International Conference on Biomolecular Engineering

**Group:** General Submissions

**Session:** Technical Session 4: Self-assembly

**Time:** Wednesday, January 8, 2020 - 2:40pm-3:05pm

## Authors:

Care, A., Macquarie University
Diaz, D., Macquarie University
Sunna, A., Macquarie University
Vidal Asensio, X., Macquarie University

Photodynamic therapy (PDT) is a selective and non-invasive cancer treatment. To destroy tumour cells, PDT relies on photosensitizers that are activated by light to convert the oxygen within tumour cells into toxic reactive oxygen species (ROS) that induce cell death. The fluorescent proteins KillerRed (KR) and mini-Singlet Oxygen Generator (mSOG) are unique biological photosensitizers that produce ROS when irradiated with light. Herein, encapsulins (ENC), a class of protein-based nanoparticles found in prokaryotes, were engineered to encapsulate KR or mSOG variants. All photosensitizer-loaded ENCs were produced in *Escherichia coli* and were ~30 nm in size, monodisperse and fluorescent. KR (Type I photosensitizer) generates ROS under green/yellow light. Upon activation with green light, KR-loaded ENC (KR-ENC) produced similar amounts of ROS as free KR, while unloaded ENC generated no ROS. We also visualised the rapid internalisation of KR-ENC by human brain cancer cells, confirming ENCâ€™s feasibility as a nanocarrier for functional biological photosensitizers. mSOG (Type II photosensitizer) produces singlet oxygen ($^1O_2$) upon blue light irradiation. mSOG variants, mSOG-1 and mSOG-2, previously engineered for enhanced $^1O_2$ generation were loaded into ENC. All mSOG-loaded ENCs produced $^1O_2$ under blue light activation, with a mSOG-1-loaded ENC variant (mSOG-1-ENC) shown to be the most effective. Next, we evaluated the PDT killing effect of mSOG-1-ENC in a cell model of lung cancer. mSOG-1-ENC displayed no cytotoxicity in the dark, however, when activated with blue light, it caused a ~25% reduction in cancer cell viability. These results show light-activated mSOG-1-ENCâ€™s photosensitizing capacity and its ability to mediate PDT.

# Lipid-Peptide Co-Assemblies for Multifunctional Biointerfaces

**Conference:** International Conference Biomolecular Engineering ICBE

**Year:** 2020

**Proceeding:** ICBE Asia 2020 - 10th International Conference on Biomolecular Engineering

**Group:** General Submissions

**Session:** Technical Session 4: Self-assembly

**Time:** Wednesday, January 8, 2020 - 2:15pm-2:40pm

## Authors:

Salentinig, S., University of Fribourg
Zabara, M., Empa
Gontsarik, M., Empa

Amphiphilic lipids and peptides can self-assemble to form a diversity of structures with functionality on multiple length scales. Directing the self-assembly and structure formation provides opportunities for the design of functional biointerfaces for various applications including medical therapies and food products.[1]

This presentation demonstrates the co-assembly of selected food lipids with peptides into functional biointerfaces that kill bacteria and promote wound healing.[2] We further show that external variables including pH, humidity, enzymes and temperature can be used to tailor the biointerface morphology and trigger the antibacterial activity on demand.[3,4,5] These materials are sustainable, biodegradable and food grade. The mechanisms at play underlying the self-assembly structure formation and its dynamic transformations with external triggers are studied using highly contemporary biophysical and computational methods including in situ time-resolved synchrotron SAXS and GISAXS, bio-AFM, cryo-TEM and online confocal Raman microscopy. Additional antibacterial assays on a broad range of both gram negative and positive bacteria strains, together with toxicity assays bridge the colloidal structure and composition to the biological activity.

The detailed insights into dynamic self-assembly in bio-mimetic materials and their characterisation on multiple length scales may provide essential knowledge for the comprehensive design of functional biointerfaces in form of nanoparticles and coatings.


1. Salentinig S., *Curr. Opin. Coll. Interf. Sci.* **2019**, *39*, 190-201.
2. Zabara M., et al. *Adv. Funct. Mat.*, **2019**, 29, 1904007.
3. Gontsarik M., et al. *ACS Appl. Mater. Interfaces*, **2019**, 2821-2829.
4. Gontsarik M., et al. *Biomat. Sci.* **2018**, *6*, 803-812.
5. Salentinig S., et al. *Phys Chem Chem Phys.* **2018**, *20*, 21903-21909.

# Plenary Talk: Cell Surface Topography: How Protein Size can Alter Organization and Signaling at Cell-Cell Interfaces

**Conference:** International Conference Biomolecular Engineering ICBE

**Year:** 2020

**Proceeding:** ICBE Asia 2020 - 10th International Conference on Biomolecular Engineering

**Group:** General Submissions

**Session:** Plenary Lecture III: Biomolecular Analytics and Characterizations

**Time:** Thursday, January 9, 2020 - 9:30am-9:55am

## Authors:

Fletcher, D. A., University of California, Berkeley

Membrane interfaces formed at junctions between cells are often associated with characteristic patterns of membrane protein organization, such as in epithelial tissues and between cells of the immune system. The size of cell surface proteins is emerging as a critical property that can directly affect cell-cell interface formation and contribute to spatial arrangement of membrane proteins at junctions, as well as their downstream signaling. This talk will describe a new method for characterizing cell surface protein size that enables nanometer-scale height measurements. With this information, we use in vitro systems based on giant unilamellar vesicles and live immune cells to show that fluid membrane interfaces linked by adhesion proteins can drive segregation of non-adhesive proteins, directly affecting macrophage signaling during receptor-mediated target recognition. Results from these studies support a model in which cell surface protein size plays a key role in mediating cell-cell communication and function.

# Explore the Adhesion Mechanisms of Mussel Adhesive Proteins

**Conference:** International Conference Biomolecular Engineering ICBE

**Year:** 2020

**Proceeding:** ICBE Asia 2020 - 10th International Conference on Biomolecular Engineering

**Group:** General Submissions

**Session:** Technical Session 5: Analytics

**Time:** Thursday, January 9, 2020 - 11:05am-11:30am

## Authors:

Yu, J., Nanyang Technological University, Singapore

The rapid and robust adhesion of marine mussels to diverse solid surfaces in wet environments is mediated by the secreted mussel adhesive proteins which are abundant in a catecholic amino acid, *L*-3,4-dihydroxyphenylalanine (Dopa). Over the last two decades, enormous efforts have been devoted to the development of synthetic mussel-inspired adhesives with water-resistant adhesion and cohesion properties by modifying polymer systems with Dopa and its analogues. Using the surface forces apparatus (SFA), we systematically explored the adhesion mechanism of various mussel foot proteins. Our SFA results show that mussels achieve strong interfacial binding via balancing a variety of covalent and noncovalent interactions including oxidative cross-linking, electrostatic interaction, metal-catechol coordination, hydrogen bonding, hydrophobic interactions and Ï€â€“Ï€/cation-Ï€ interactions. The SFA can also measure the adhesion and interfacial tension of complex coacervates formed by mussel foot proteins (Mfps). The insights obtained from the understanding of the interaction mechanisms between individual Mfp and disparate substrates provide critical guidance for the design of next-generation wet adhesive materials.

# Invited Talk: Zika Virus and its Antibody Complex Structures

**Conference:** International Conference Biomolecular Engineering ICBE

**Year:** 2020

**Proceeding:** ICBE Asia 2020 - 10th International Conference on Biomolecular Engineering

**Group:** General Submissions

**Session:** Technical Session 5: Analytics

**Time:** Thursday, January 9, 2020 - 10:15am-10:40am

## Authors:

Lok, S. M., Duke University-National University of Singapore

ZIKV is a member of the family *flaviviridae*, other members in the same family are West Nile, dengue, yellow fever viruses. ZIKV was previously thought to cause only a mild febrile disease characterized by fever, rash, joint pain and red eye, in only 20% of the infected humans. However, in the recent outbreaks in South America and French Polynesia, reports of an increased incidence of Guillain-BarrÃ© syndrome (GBS) in adults and microcephaly in foetuses associated with the disease sounded the global alarm. In addition to spreading by the bite of an infected mosquito, ZIKV can also be spread by sexual intercourse suggesting that the transmission could be faster than DENV. Therefore, there is an immediate need to develop therapeutics and vaccines. To produce a safe vaccine and therapeutics, more basic understanding of ZIKV is required. Here we will discuss the high resolution ZIKV structure and the development of antibody therapeutics and vaccine development using structural information obtained by cryo-electron microscopy.

# Time-Resolved Small-Angle X-Ray Scattering for Monitoring the Biomolecular Assembly Process

**Conference:** International Conference Biomolecular Engineering ICBE

**Year:** 2020

**Proceeding:** ICBE Asia 2020 - 10th International Conference on Biomolecular Engineering

**Group:** General Submissions

**Session:** Technical Session 5: Analytics

**Time:** Thursday, January 9, 2020 - 10:40am-11:05am

## Authors:

Ikeguchi, M., Soka University

The small-angle X-ray scattering (SAXS) is a useful technique to know the assembly state of biomolecules. We can obtain the molecular weight of the assembled complex from the forward scattering intensity, that is, we can know how many molecules are included in the complex. The scattering profile also provides the information about the shape of the complex. The development of a photon-counting PILATUS detector made it possible to observe two-dimensional scattering images at every few milliseconds. Using this detector and a stopped-flow apparatus, we followed the assembly reactions of *Escherichia coli*ferritin A (EcFtnA) and its mutants by the time-resolved SAXS (TR-SAXS) method. EcFtnA forms a cage-like structure that consists of 24 identical subunits and dissociates into dimers at acidic pH. The dimer maintains native-like secondary and tertiary structures and is able to reassemble into a 24-mer when the pH is increased. The reassembly reaction was induced by pH jump and reassembly was followed by TR-SAXS. Time-dependent changes in the forward scattering intensity and in the radius of gyration suggested the existence of a significant population of intermediate oligomers during the assembly reaction. The initial reaction was a mixture of second- and third-order reactions (formation of tetramers and hexamers) from the protein-concentration dependence of the initial velocity. The time-dependent change in SAXS profile could be explained by a simple model in which only tetramers, hexamers, and dodecamers were considered as intermediate.

# Converting Fatty Acids to Alkanes with Blue Light

**Conference:** International Conference Biomolecular Engineering ICBE

**Year:** 2020

**Proceeding:** ICBE Asia 2020 - 10th International Conference on Biomolecular Engineering

**Group:** General Submissions

**Session:** Technical Session 6: Applications

**Time:** Thursday, January 9, 2020 - 12:05pm

## Authors:

Lake, J., University of California, Berkeley
Curran, S., University of California, Berkeley

The production of fatty-acid derived fuels and chemicals by microbial metabolism offers a renewable alternative to petroleum-based fuels on the market. Microbial hydrocarbon biosynthesis in particular is a target for metabolic engineering efforts due the potential use of alkanes and alkenes as drop-in transportation fuels. Here, we explore the development of an *Escherichia coli* strain engineered for increased fatty acid production and conversion of fatty acids to alkanes by a light-driven enzyme. This photoenzyme, named fatty acid photodecarboxylase (FAP), is native to microalgae *Chlorella variabilis* and has been shown to directly decarboxylate fatty acids to their corresponding $C_{n-1}$ alkanes in a reaction catalyzed by blue light. By deleting the *fadD* and *fadE* genes responsible for fatty acid degradation and expressing FadR and â€˜TesA proteins for increased free fatty acids, we were able to show increased alkane production by the FAP enzyme *in vivo*. Our findings suggest that the FAP enzyme could be useful for the production of biodiesel-like alkanes and further investigation into other light-regulated metabolic pathways is warranted.

# Ip Is Decreased Melanogenesis Using Inhibition of PKA

**Conference:** International Conference Biomolecular Engineering ICBE

**Year:** 2020

**Proceeding:** ICBE Asia 2020 - 10th International Conference on Biomolecular Engineering

**Group:** General Submissions

**Session:** Technical Session 6: Applications

**Time:** Thursday, January 9, 2020 - 12:05pm

## Authors:

Cha, H. J., Osan university

Melanogenesis is essential UV protection mechanism in skin. However, excessive melanogenesis causes hyperpigmentation deceases, such as a melasma, freckle, etc. Therefore, through depigmentation agents, it is necessary to regulate melanogenesis. In this study, we identified noble whitening agent and their molecular mechanism. Firstly, we show that IP led to decrease melanin contents in a dose dependent manner in B16F10. The decrease of melanin synthesis was occurred by IP mediated decreasing Tyrosinase activity and expression. In general, Tyroainase expression is regulated by Microphthalmia-associated transcription factor (MITF) in a-MSH treated B16F10. Thus, we identified change of MITF activity in IP treated B16F10. IP down-regulated MITF activity and also MITF expression in a-MSH treated B16F10. Next, we identified CREB activity and phosphorylation in IP-treated B16F10, because CREB activity is implicated in MITF expression in a-MSH mediated melanogenesis. As shown our result, IP decreased CREB activity and phosphorylation in B16F10. Additionally, as CREB is phosphorylated by PKA, we showed PKA activity using phosphor-PKA. Phosphor-PKA was up-regulated by IP in B16F10. Overall, IP regulates melanin synthesis using regulation of PKA-CREB-MITF-Tyrosinase axis.

Acknowledgments

This research was supported by Basic Science Research Program through the National Research Foundation of Korea (NRF) funded by the Ministry of Education (2017R1D1A1B03028380).

# "Parallel Metabolic Pathway Engineering" for the Bioproduction of Shikimate Pathway Derivatives

**Conference:** International Conference Biomolecular Engineering ICBE

**Year:** 2020

**Proceeding:** ICBE Asia 2020 - 10th International Conference on Biomolecular Engineering

**Group:** Poster Session

**Session:** Poster Session

## Authors:

Fujiwara, R., Kobe university
Tanaka, T., Kobe University
Kondo, A., Kobe University
Noda, S., Riken

Glucose and xylose are the major components of lignocellulosic biomass, and the effective use of both sugars contributes to improving the efficiency of bioproduction. Here we propose a method termed â€œparallel metabolic pathway engineeringâ€ (PMPE) for producing shikimate pathway derivatives from glucoseâ€“xylose co-substrate. In this method, we seek to use glucose mainly for target chemical production and xylose for supplying essential metabolites for cell growth. Glycolysis and the pentose phosphate pathway (PPP) are completely separated from the tricarboxylic acid (TCA) cycle. To recover cell growth, a xylose catabolic pathway that directly flows into the TCA cycle is introduced. We focused on the Dahms pathway which is a xylose catabolic pathway in*Caulobacter crescentus*, an oligotrophic bacterium. The Dahms pathway directly produces pyruvate and glyoxylate from xylose without glycolysis and PPP. To prove our concept, we produced *cis,cis*-muconic acid (MA) using the PMPE *Escherichia coli* strain. MA, a shikimate pathway derivative, is a valuable compound that is a precursor of important chemical compounds such as adipic acid and terephthalic acid, and its production has recently attracted attention. We modified the metabolic pathway of *E. coli* so that only glucose could be used for MA production, and we introduced the Dahms pathway to restore cell growth. As a result, we produce 4.09 g/L of MA using the PMPE *E. coli* strain with a high yield (0.31 g/g of glucose). The â€œPMPEâ€ strategy will contribute to the development of clean processes for producing various valuable chemicals from lignocellulosic resources.

# CXCR4 Regulates Branching Morphogenesis in Developing Epithelial Organ

**Conference:** International Conference Biomolecular Engineering ICBE

**Year:** 2020

**Proceeding:** ICBE Asia 2020 - 10th International Conference on Biomolecular Engineering

**Group:** Poster Session

**Session:** Poster Session

## Authors:

Kim, J., Seoul National University

The CXC Chemokine receptor type 4 (CXCR4), a member of 7-transmembrane receptor family, is known to facilitate migration of immune cells as well as stem cell homeostasis. Not only that, CXCR4 has gained extensive attention due to its ambivalent augmentation effects on the development of organs. Many studies have been conducted on the contributions of CXCR4 to organogenesis and, specifically, to neurogenesis and angiogenesis. However, the function of CXCR4 in the early branching stages of embryo submandibular gland is yet to be explored. In this study, we investigated the relations between CXCR4 and glandular branching morphogenesis. AMD3100, a potent CXCR4 inhibitor, was applied to embryo submandibular glands extracted from E13 and E14 embryos. E14 eSMGs showed a retarded growth compared to control group whereas the branching and expansion were nearly abolished in E13 eSMGs. Both control and AMD3100-treated groups, however, showed no significant difference in their expressions of cleaved caspase-3. Analysis of mRNA sequencing data revealed considerable increase in the expressions of acinar and ductal progenitor as well as of differentiation marker genes. Collectively, our results show that CXCR4 regulates the differentiation of acinar and ductal cells in glandular branching morphogenesis.

# Development of Computer-Aided Rational Protein Engineering Toolkit (CARPET)

ICBE Asia

**Conference:** International Conference Biomolecular Engineering ICBE

**Year:** 2020

**Proceeding:** ICBE Asia 2020 - 10th International Conference on Biomolecular Engineering

**Group:** Poster Session

**Session:** Poster Session

## Authors:

Jeon, H. N., Yonsei University
Lim, H., Bioinformatics and Molecular Design Research Center (BMDRC)
Hwang, S., Yonsei University
Kim, J., Bioinformatics and Molecular Design Research Center (BMDRC)
No, K. T., Yonsei University

Protein engineering is varying the structure of a protein to develop valuable proteins. Protein side-chain mutation is fundamental to protein engineering processes, but the astronomical number of all possible protein sequences makes protein engineering difficult. Thus, computer-aided protein engineering is required to explore all possible protein sequence space and to guide the effective protein design. A number of different approaches have been suggested and applied to prediction of protein properties, such as enzyme reactivity, binding affinity, substrate/product selectivity, thermal/pH stability, and solubility, through sequence-based machine learning algorithms, knowledge-based potential functions, all-atom molecular mechanics based calculations, and all-atom quantum mechanics based calculations. Here, we suggest a new Computer-Aided Rational Protein Engineering Toolkit (CARPET) which predicts protein properties via sequence-based algorithms and structure-based algorithms, and helps rational enzyme modification. In CARPET, it starts from phylogenetic analysis via structural superimposition for inter-species comparison. The good score mutations are selected from the suggested mutation pool based on prediction models of thermal stability and solubility. Free Energy Perturbation (FEP) and Thermodynamic Integration (TI) are used to calculate more accurate thermal/pH stability. Finally, fragment molecular orbital (FMO) method is used to calculate calculate residue-residue interactions in enzyme to predict protein properties. Through collaborative manners in our new platform, we can analyze the experimental mutation results, design putative mutation pool, predict the properties of suggested mutation, and finally pick out the best-score mutations.

# Differences in Loop Motions and Dynamics in the Catalytic Domains of GH Family 19 Chitinases Revealed By MD Simulation

**Conference:** International Conference Biomolecular Engineering ICBE

**Year:** 2020

**Proceeding:** ICBE Asia 2020 - 10th International Conference on Biomolecular Engineering

**Group:** Poster Session

**Session:** Poster Session

## Authors:

Oyeleye, A., 1Enzyme and Microbial Technology Research Centre
Chor Leow, A. T., Universiti Putra Malaysia
Saidi, N. B., Universiti Putra Malaysia
Mohd Yahaya, N., Universiti Putra Malaysia

Plants produce GH family 19 chitinases as a means of defence against invading fungal pathogens. This family of chitinases has been identified and characterized in a few bacterial species including *Streptomyces griseus* and *Streptomyces coelicolor*. Reports have shown that bacterial chitinases require a chitin binding domain for their inhibitory properties against fungi while their plant counterparts in classes I and II of GH family 19, do not require same. Here, we have identified some distinct loop motions in a 20 ns molecular dynamics simulation that suggests that some conformational changes may be triggered by loops in the catalytic domain of plant chitinases which may be essential for recognition and binding to complex or crystalline chitin substrates such as is present in fungal cell walls. In the bacterial chitinases however, the catalytic domain undergoes fewer flexible movements. We propose that such minimal flexibility which might be optimal for binding to chitooligosaccharides as revealed in previous reports, is not sufficient for binding to crystalline chitin. Currently, studies are underway to experiment the influence of redesigning the catalytic domain of *Streptomyces griseus* chitinase C (ChiC) towards enhancing flexibility in the catalytic domain and subsequently improving its hydrolytic and antifungal properties.

# Metabolic Engineering to Improve 1,5-Diaminopentane Production from Cellobiose Using Bgl-Secreting Corynebacterium glutamicum

**Conference:** International Conference Biomolecular Engineering ICBE

**Year:** 2020

**Proceeding:** ICBE Asia 2020 - 10th International Conference on Biomolecular Engineering

**Group:** Poster Session

**Session:** Poster Session

## Authors:

Tanaka, T., Kobe University
Kondo, A., Kobe University

Microbial production of 1,5-diaminopentane from renewable feedstock is a promising and sustainable approach for the production of polyamides. We constructed a Î$^{2}$-glucosidase (BGL)-secreting *Corynebacterium glutamicum* and successfully used this strain to produce1,5-diaminopentane from cellobiose and glucose. First, *C. glutamicum* was metabolically engineered to produce l-lysine, a direct precursor of 1,5-diaminopentane, followed by the co-expression of l-lysine decarboxylase and BGL derived from *Escherichia coli* and *Thermobifida fusca* YX (Tfu0937), respectively. This new engineered *C. glutamicum* strain produced 27 g/L of 1,5-diaminopentane from cellobiose in CGXII minimal medium using fed-batch cultivation. The yield of 1,5-diaminopentane was 0.43 g/g glucose. These results demonstrate the feasibility of1,5-diaminopentane production from cellobiose or cellooligosaccharides using an engineered *C. glutamicum* strain.

# Microbial Platform to Synthesize Valuable Bulk Chemicals in Escherichia coli

**Conference:** International Conference Biomolecular Engineering ICBE

**Year:** 2020

**Proceeding:** ICBE Asia 2020 - 10th International Conference on Biomolecular Engineering

**Group:** Poster Session

**Session:** Poster Session

## Authors:

Noda, S., Riken

A synthetic metabolic pathway suitable for the production of chorismate derivatives was designed in *Escherichia coli*. An L-phenylalanine-overproducing *E. coli* strain was engineered to enhance the availability of phosphoenolpyruvate (PEP), which is a key precursor in the biosynthesis of aromatic compounds in microbes. Two major reactions converting PEP to pyruvate were inactivated. Using this modified *E.coli* as a base strain, we tested our system by carrying out the production of salicylate, a high-demand aromatic chemical. The titer of salicylate reached 11.5 g/L in batch culture after 48 h cultivation in a 1-liter jar fermentor, and the yield from glucose as the sole carbon source exceeded 40% (mol/mol). In this test case, we found that pyruvate was synthesized primarily via salicylate formation and the reaction converting oxaloacetate to pyruvate. In order to demonstrate the generality of our designed strain, we employed this platform for the production of each of 7 different chorismate derivatives. Each of these industrially important chemicals was successfully produced to levels of 1-3 g/L in test tube-scale culture. In addition, by extending chorismate pathway, we successfully achieved maleate production, which is one of significant dicarboxylic acids as well as succinate and malate. A novel synthetic pathway of maleate was constructed in our base strain, and the production reached 7.1 g/L. This is the first report about maleate production using genetically engineered micro-organisms.

# Polarisome Scaffolder Spa2-Mediated Macromolecular Condensation of Aip5 for Actin Polymerization

**Conference:** International Conference Biomolecular Engineering ICBE

**Year:** 2020

**Proceeding:** ICBE Asia 2020 - 10th International Conference on Biomolecular Engineering

**Group:** Poster Session

**Session:** Poster Session

## Authors:

Xie, Y., Nanyang Technological University
Sun, J., Nanyang Technological University
Han, X., Nanyang Technological University
Wunder, A. T., Nanyang Technological University
Gao, Y., Nanyang Technological University
Miao, Y., Nanyang Technological University

Polarized fungal growth requires the highly-coordinated assembly of actin cable aligning along the mother-daughter axis. For directional actin polymerization during polarized growth, actin cable nucleator formin and formin nucleation promoting factors maintain a condensed-zone at the bud tip. A macromolecular complex called polarisome plays essential roles in regulating these actin-binding proteins in a spatial-temporal manner throughout the cell cycle progression. However, the protein constituents in the polarisome and their intra- or inter-molecular interactions for the dynamic regulation of protein functions in actin assembly and polarized cell growth remain elusive. Recently, we have identified a previously uncharacterized actin-interacting protein 5 (Aip5), which directly interacts with formin and Spa2, the scaffolder of the polarisome complex. Through the cell biology, biochemical, and structural biology studies, we have revealed that by interacting with formin, Aip5, independently or synergistically nucleates actin filament assembly. Furthermore, using both in vivo cell imaging and in vitro protein reconstitution experiments, we found that the intrinsically disordered region of Aip5 plays essential roles in regulating the protein function through phase separation mechanism, in a Spa2- and actin-dependent manner, during stress adaptation. The macromolecular condensation of Aip5 provides a few new mechanisms underlying the regulation of actin polymerization during polarized fungal growth or stress adaptation.

# Salt-Induced Assembly Kinetics of Archaeoglobus Fulgidus Ferritin Monitored By Time-Resolved Small-Angle X-Ray Scattering

**Conference:** International Conference Biomolecular Engineering ICBE

**Year:** 2020

**Proceeding:** ICBE Asia 2020 - 10th International Conference on Biomolecular Engineering

**Group:** Poster Session

**Session:** Poster Session

## Authors:

Kuwata, T., Soka Univerisity
Sato, D., Soka Univerisity
Lim, S., Nanyang Technological University
Ikeguchi, M., Soka University
Kumar, A., Nanyang Technological University
Law, B., Nanyang Technological University
Tan, T., Nanyang Technological University

Ferritin is a protein cage composed of 24 identical subunits with octahedral 4-3-2 symmetry. *Archaoglobus fulgidus* ferritin (AfFtnWT) has a unique â€œopen structureâ€ shifting from 4-fold symmetry to tetrahedral 2-3 symmetry forming 4 triangular pores with 52Ã… edges in the shell. It has been shown that the replacements of Lys-150 and Arg-151 to alanines dramatically change the assembly of AfFtn. The mutant K150A/R151A (AfFtnAA) assembled into a conventional octahedral cage, in which Ala-150 and Arg-151 were located at 4-fold symmetry axis. Furthermore, AfFtnWT and AfFtnAA exist as subunit dimers having native-like secondary and tertiary structures at low salt concentrations. They can reversibly reassemble into respective 24mers when salt concentration is increased. In this study, the assembly reactions of AfFtnWT and AfFtnAA were induced by NaCl-concentration jump and monitored by time-resolved small-angle x-ray scattering (TR-SAXS). The results indicated that the assembly rate of AfFtnAA is 25 times faster than that of AfFtnWT of 445mM NaCl. Although the detailed analysis of the assembly reaction was hampered by the unknown large aggregate formation at higher protein concentrations under the low salt conditions, the kinetic curve of AfFtnAA was similar to that of*Escherichia coli* ferritin (EcFtnA), indicating that AfFtnAA assembles into 24mer via the mechanism similar to that previously elucidated for EcFtnA.

# Selection of Peptides Inhibiting p53 Function

**Conference:** International Conference Biomolecular Engineering ICBE

**Year:** 2020

**Proceeding:** ICBE Asia 2020 - 10th International Conference on Biomolecular Engineering

**Group:** Poster Session

**Session:** Poster Session

## Authors:

Chee, S., p53LAB (A*STAR)
Jia Wei, S., p53LAB (A*STAR)
Ghadessy, F., p53LAB (A*STAR)
Ting Xiang, L., p53LAB (A*STAR)
Yen-Chu Lin, J., p53LAB (A*STAR)

As â€œGuardian of the Genomeâ€, the p53 tumour suppressor plays a major role in determining cell fate. Mutant forms of p53 are commonly seen in cancer, and can drive tumour development through gain of function activities. The wild-type state of p53 is responsible for chemotherapy-associated side effects. Agents modulating p53 function are therefore of significant interest. We have used yeast-display to select for Avian Pancreatic Peptide (APP) derivatives that bind specifically to p53. The lead selectant shows on-target association with p53 as measured by pull-down assays, both in vitro and in cells. Furthermore, it is able to inhibit the activity of wild-type p53 in cells. This peptide is a useful tool for disrupting p53 function and is currently being further engineered to enhance its function.

# Solar Spectrum Characterization, Impact on Human Health and Mitigation Using Innovative Materials

**Conference:** International Conference Biomolecular Engineering ICBE

**Year:** 2020

**Proceeding:** ICBE Asia 2020 - 10th International Conference on Biomolecular Engineering

**Group:** Poster Session

**Session:** Poster Session

## Authors:

Abu Alhawa, D., Texas A&M University at Qatar
Roustazadeh, L., Texas A&M University at Qatar
Kokosimos, K., Texas A&M University at Qatar
Grati, A., Texas A&M University at Qatar
Abousrafa, A., Texas A&M University at Qatar

Solar radiation, depending on the degree of exposure, is a key factor on the manufacturing of vitamin-D, especially the UVB portion of light (290-320 nm). However, excessive exposure to solar UV-B irradiance has detrimental effects to human health, since it is a genotoxic stressor, and cause unrepairable DNA damage, skin cancer etc. Surprisingly, Vitamin-D deficiency and metabolic disorders are common in Middle Eastern countries (especially Qatar) and have been linked to inadequate exposure to sunlight, regardless of the more than 9-hours average sunshine per year. Beyond the biological complexity of the above phenomena other environmental factors, as well, have a significant impact.

Efforts in this study focus on the quantification of the threshold exposure time needed in Qatar to produce vitamin-D. In other words, to create a model relating the threshold UV exposure time and the production of vitamin-D based on literature and local data. To achieve this, we assembled an experimental setup to characterize solar irradiance and spectral composition (intensity per wavelength). Measurements collected from a pyranometer and spectrometer were combined to obtain graphs of the energy and wavelength of spectrum throughout the day. In parallel, by conducting a literature review of the implications of UV irradiation, a list of different biomarkers was identified as causing significant health impacts, such as cyclobutane pyrimidine dimers (CPD) and 8-oxo-dG. Here, we present the preliminary results of our work and the python-model we developed to utilize local atmospheric data (through AERONET), our measurements, and a literature Vitamin-D synthesis model.

# Studying Performance of Microbial Fuel Cell Inoculated with Shewanella Oneidensis MR-1 for Copper Recovery

**Conference:** International Conference Biomolecular Engineering ICBE

**Year:** 2020

**Proceeding:** ICBE Asia 2020 - 10th International Conference on Biomolecular Engineering

**Group:** Poster Session

**Session:** Poster Session

## Authors:

Bhagwat, A., University of British Columbia (Vancouver)

Conventional methods of extracting metals involve fragmentation and processing of ores through toxic chemicals. Along with the high cost and energy for processing low grade ores, dealing with immense amount of waste rock is a formidable challenge for the mining industry. Copper at high concentrations becomes anthropogenic and can be particualarly be found in acid mine drainage (AMD). Bio-electrochemical systems provide low operating costs and are a sustainable alternative for many industrial applications. Microbial fuel cells (MFC) have advanced from electricity production to wastewater treatment. Species of bacteria known as exoelectrogen can transfer electrons from their extracellular structures such as pili and flavins. These bacteria have a very well-defined metabolic pathways leading to terminal electron acceptors. Shewanella Oneidensis MR-1 has been one of the most widely studied exoelectrogen. As part of my undergraduate thesis, I am studying the kinetics of electron transport in a two-chamber microbial fuel cell inoculated with Shewanella Oneidensis MR-1 under lactate substrate and copper sulphate solution at cathode. The study also explores recovery of copper at carbon neutral cathode. The maximum power density achieved by the cell was 13 $mW/m^2$ with 40% removal of copper from catholyte. Furthermore, I am working on genetically engineering the strain by overexpressing genes encoding for proteins playing pivotal role in electron transfer to boost extracellular electron transfer, thereby generating higher voltage and increasing rate of copper recovery. I look forward to more development on applications of MFCs and inspire research in scaling up bio-electrochemical systems to meet industry demands.

# The Comparison of MSC Secretome Effects on Corneal Wound Healing Depending on the Cell Culture Environments

**Conference:** International Conference Biomolecular Engineering ICBE

**Year:** 2020

**Proceeding:** ICBE Asia 2020 - 10th International Conference on Biomolecular Engineering

**Group:** Poster Session

**Session:** Poster Session

## Authors:

Lee, H. J., Gachon University

The therapeutic effects of secreted factors (secretome) produced by human mesenchymal stem cells (MSCs) were evaluated depending on the cell culture environments. The cells were cultured on standard 2D culture conditions and 3D electrospun fiber scaffolds, and the secretomes were collected. The goal was to determine whether culturing MSCs within a 3D, extracellular matrix-like environment would improve the therapeutic.

The secretome of MSCs cultured on 2D substrates and 3D electrospun fiber scaffolds showed substantial compositional differences. The relative amounts of factors such as HGF and ICAM-1 were increased over 5 times in 3D cultures compared to 2D cultures. Cell viability of corneal fibroblasts sustained metabolic activity for 6 days when co-cultured with MSCs grown on the electrospun fiber scaffolds in the absence of serum. A scratch-based wound healing assay showed 95% closure at 48 hours in corneal fibroblast monolayers co-cultured with MSCs grown on electrospun fibers, while the control group only exhibited 50% closure at the same time point. When electrospun fibers with seeded MSCs were applied to an *ex vivo* rabbit corneal organ culture system, it promoted re-epithelialization. Immunostaining of the corneas showed that the expression of alpha-SMA was lower compared to the injured cornea without treatment. In addition, keratan sulfate expression was higher in the treated group compared to the untreated.

MSCs cultured on three-dimensional, electrospun fiber scaffolds facilitate wound healing effects on corneal fibroblasts *in vitro* and on explanted corneas in an organ culture model through differential secretome profiles compared to MSCs cultured on 2D substrates.

# Theasaponin E1 Attenuate Amyloid Beta(A?) Proteins and Hyperphosphorylated Tau in Sweapp N2a and Shy-5Y Cells By Down or up-Regulation of the Associated Signaling Molecules and Enzymes

**Conference:** International Conference Biomolecular Engineering ICBE

**Year:** 2020

**Proceeding:** ICBE Asia 2020 - 10th International Conference on Biomolecular Engineering

**Group:** Poster Session

**Session:** Poster Session

## Authors:

KIM, J. D., Chonnam Natational University
Khan, M. I., Chonnam Natational University
KIM, M. Y., Chonnam Natational University

In present study we investigated the therapeutic effects of green tea seed isolated theasaponin E1on attenuating Alzheimer by ameliorating the AÎ² and tau level in SweAPP N2a cells and SHY-5Ycells. For reducing AÎ², the inhibition of Î², Î³-secretases and acetylcholinesterase and activation of Î±-secretase (ADAM10), neprilysin and insulin degrading enzyme (IDE) was investigated. In order to determine the inhibitory or activating effects of theasaponin E1 on tau, we targeted the signaling proteins involved in tau phosphorylation and dephosphorylation (Gsk-3Î², Cdk5, MARK, Jnk, Akt, CaMKII AMPK/ERK, PP1, PP2-A and PP2-B). investigation was done with fluorometric assays, RT-PCR, Western blotting and ELISA respectively. Theasponin E1 was extracted and purified from green tea seed extract via HPLC. Cultured cells after treating with various concentrations were then processed for extraction of RNA, proteins and cell lysate which was then used for fluorometric assays, RT-PCR, Western blotting and ELISA respectively. Gen specific primers were used for amplification of the target genes and specific antibody for each terget were used for expression and quantification of the proteins. Our results demonstrated that theasaponin E1 significantly reduced AÎ² concentration by activation of Î±-Secretase (over expression of ADAM10) and neprilysin. Activities of Î² secretase and Î³-secretase were reduced in dose-dependent manner due to down-regulation of BACE1, Presenilin (PS1) and Nicastrin (NCT) respectively. NCT inhibition was lower as compared to PS1 also theasaponin E1 significantly reduce the activities of acetylcholinesterase. Theasaponin E1 effectively inhibit tau phosphorylation in dose dependent manner via downregulating Gsk-3Î², Cdk5, MARK, Jnk, CaMKII.

# AUTHOR INDEX


Abousrafa, Asma....36
Alhawa, Dima Abu....36
Asensio, Xavier Vidal....18
Bagh, Sangram....16, 17
Baltz, Morgan R....7
Belcher, Donald....3
Bhagwat, Arjun....37
Birch, Mark A....2
Bonnerjee, Deepro....17
Buecheler, Jakob....6
Care, Andrew....18
Cha, Hwa Jun....26
Chee, Sharon....35
Chen, Xixian....14
Chung, Muen....12
Curran, Samuel....25
Delisa, Matthew P....7
Diaz, Dennis....18
Dura, Gema....2
Fang, Mengjun....4
Fletcher, Daniel A....21
Freemont, Paul S....15
Fujiwara, Ryosuke....27
Fulton, David A....2
Gao, Yonggui....33
Ghadessy, Farid....35
Gontsarik, Mark....19
Grati, Amani....36
Han, Xiao....33
Howarth, Mark....5
Huang, Lei....4
Hwang, Sungbo....29
Ikeguchi, Masamichi....24, 34
Jeon, Hyeon Nae....29
Jiang, Allen....7
Jungbauer, Alois....9
Khan, Muhammad Imran....39
Kim, Jong Deog....39
Kim, Jongwan....29
Kim, Junchul....28
Kim, Min Yong....39
Kim, Young-Kee....13
Kokosimos, Konstantinos....36
Kondo, Akihiko....27, 31
Kumar, Ambrish....34
Kuwata, Takumi....34
Lake, Julie....25
Lakey, Jeremy H....2
Law, Boyce....34
Lee, Hyeon Cheol....7
Lee, Hyun Jong....38
Leow, Adam Thean Chor....30
Lim, Hocheol....29
Lim, Sierin....34
Lin, Jimmy Yen-Chu....35
Lok, Shee-Mei....23
Meksiriporn, Bunyarit....7
Miao, Yansong....33
No, Kyoung Tai....29
Noda, Shuhei....27, 32
Oyeleye, Ayokunmi....30
Park, Jisoo....10
Perkins, Neil D....2
Peters, Daniel T....2
Roustazadeh, Laya....36
Saidi, Noor Baity....30
Salentinig, Stefan....19
Sarkar, Kathakali....16
Sato, Daisuke....34
Shim, Jeung Yeop....10
Shukal, Sudha....14
Stephens, Erin....7
Stupp, Samuel....11
Sun, Jialin....33
Sunna, Anwar....18
Tam, James P....1
Tan, Tabitha....34
Tanaka, Tsutomu....27, 31
Thi, Hien Nguyen....13
Waller, Helen....2
Wang, Haijiao....4
Waraho, Dujduan....7
Wei, Siau Jia....35
Wunder, Alma T....33
Xiang, Lim Ting....35
Xie, Ying....33
Xu, Zhinan....4
Yahaya, Normi Mohd....30
Yemm, Adrian....2
Yoo, Tae Hyeon....10
Yu, Jing....22
Zabara, Mahsa....19
Zhang, Congqiang....14